OBSERVATIONS

VOL. 1

PERIODICITY

OBSERVATIONS ON W.D. GANN

VOL. 1

PERIODICITY

AWODELE

BEKH

UNION, KY

BEKH, LLC
UNION, KY
BEKHLLC@OUTLOOK.COM

ISBN-10: 0692378693
ISBN-13: 978-0692378694

Preface

It was in February of 1997 when I met my most influential spiritual teacher. A friend took me to see him for the first time during one of my breaks from classes. On the first visit he divined for me using an oracle system, performed a numerology reading on my date of birth, and told me things about myself that were quite astonishing considering that this was the first time we had ever met. I was fascinated and it sealed a friendship that has lasted ever since.

During the remaining part of 1997, which included my senior year of college and the start of graduate school, I would continue to go to his office to visit. He would always pass along books, notes, and material that he had picked up along the way to see if I resonated with any of the information. I loved it all, but will never forget the document he gave me with the title *Spiritual Development* written at the top. It was a copy of some handwritten notes dated November 2, 1975. There was no mention of the lecturer's name, but more importantly, the subject matter stimulated me like nothing else had before. I was captivated. The main theme of the lecture dealt with cycles, and how the sages would learn to live in harmony with natural cycles. It talked about how the sages used cycles to their advantage. This lit an inner fire in me to learn as much about cycles as I could.

As my interest and study of cycles grew, I obtained my first book on astrology during this time. Although this first book was on

Vedic Astrology, I have studied many forms of astrology over the years. Naturally, studying subjects such as cycles and astrology, I eventually came across the work of W.D. Gann.

William Delbert Gann was born on June 6, 1878 in Lufkin, Texas. In his promotional booklet issues in 1954, it says that he made his first trade in commodities on August 15, 1902, but his fame spread as a result of the December 1909 *Ticker and Investment Digest* magazine article written by R.D. Wyckoff, who was owner of the magazine at that time. In this article, Gann talked about "The Law of Vibration", and how it enabled him to accurately predict the points at which stocks would rise and fall. Numerous examples are given in the article where Gann predicts that a stock would not go higher or lower than a certain price. It goes on to say that in the presence of a representative of the *Ticker and Investment Digest* during the month of October 1909, Gann made 286 transactions in various stocks during 25 market days and that 240 of the 286 transactions were profitable. It said that the capital with which he operated was doubled ten times so that at the end of the month he had 1,000 percent of his original margin.

In Gann's promotional booklet entitled, *Why Money is Lost on Commodities and Stocks and How to Make Profits from 1954*, it records the following:

"1908 May 12th left Oklahoma City for New York City. August 8th made one of his greatest mathematical discoveries for predicting the trend of stocks and commodities. Started trading with a capital of $300 and made $25,000. Started another account with $130 and made $12,000 in thirty days time."

Here we have an individual that could forecast the movement of stocks months and years in advance, and his ability to do so was well documented. After reading more about him and his work, I was intrigued, and set out to learn as much as I could.

During my time studying Gann, I was fortunate to come across an e-book published by the Gann Study Group entitled, *W.D. Gann on the Law of Vibration*. In this e-book is where I first read a little known Gann article from 1919. In this article, Gann makes some predictions about the German Kaiser, Wilhelm Hohenzollern, and provides some details as to how he made his predictions. Like the document on *Spiritual Development* in 1997, I was captivated. Continued study and work on the contents of the article led to the publication of my first book entitled, *W.D. Gann: Divination By Mathematics*.

In this book, I wrote a chapter on periodicity, which was personally for me, one of my favorites in the book. In that chapter I was able to show how Gann may have derived key periods in the life of the German Kaiser, Wilhelm Hohenzollern. Interestingly, during my continued research, I found another example in Gann's work that matched the same procedure Gann may have used to derive some of the key periods in the 1919 article. Thus, the goal of this small booklet is to simply share what I found.

It has always been my intention to publish additional books to get some of the ideas that I have been working with out in the open, but I didn't have enough information on a particular topic to fill a three hundred, two hundred, or even a one hundred page book. With this in mind, I realized that I could publish small booklets in a series of volumes, each focusing on a different topic to achieve my goal. So this is what you will find in this publication, a small booklet on the topic of periodicity, which is the first volume in a series of volumes on observations I have made with respect to Gann's work. It is my hope that the reader will find something valuable within these pages to further their own research and study.

Awodele,
Union, KY

January 29, 2015

PERIODICITY

In the already mentioned *Ticker and Investment Digest* article from 1909, Gann commented on the subject of periodicity several times. In the first of these instances he states,

> "I soon began to note the periodical recurrence of the rise and fall in stocks and commodities. This led me to conclude that natural law was the basis of market movements. . . After exhaustive researches and investigations of the known sciences, I discovered that the law of vibration enabled me to accurately determine the exact points at which stocks or commodities should rise and fall within a given time."

In another section of the article, Gann says,

> "In going over the history of markets and the great mass of related statistics, it soon becomes apparent that certain laws govern the changes and variations in the value of stocks and there exists a periodic or cyclic law, which is at the back of all these movements."

Gann also says,

> "Science teaches that an original impulse of any kind finally resolves itself into a periodic or rhythmical motion; also, just as the pendulum returns again in its swing, just as the moon returns in its orbit, just as the advancing year over brings the rose of spring, so do the properties of the elements periodically recur as the weight of the atom rises."

Last, but not least, we find the statement,

> "Since all great swings or movements of the markets are cyclic they act in accordance with periodic law."

I think this last statement is very telling. Not only is he telling us that movements of the markets are cyclic, but because they are cyclic, they act in accordance with periodic law. So what is the periodic law? In a book entitled, *On The Discovery of the Periodic Law* by John A. R. Newlands, he states,

> "In an appendix to this paper (*Chemical News*, vol. x. p. 94), August 20, 1864, I announced the existence of a simple relation or law among the elements when arranged in the natural order of their atomic weights, to the effect that the eighth element, starting from a given one, was a sort of repetition of the first, or that elements belonging to the same group stood to each other in relation similar to that between the extremes of one or more octaves in music. . . In the *Chemical News*, vol. xii. pp. 83 and 94 (August 18 and 25, 1865), I published a full horizontal arrangement of the elements in order of atomic weight, and proposed to designate the simple relation existing

between them by the provisional term "law of octaves." This law has since been called by M. Mendelejeff [Mendeleev] the "periodic law.""

John Newlands was an English chemist who noted that many pairs of similar elements existed which differed by some multiple of eight in mass number, and was the first person to assign them an atomic number. When his law of octaves was printed in Chemistry News, likening this periodicity of eights to the musical scale, it was ridiculed by some of his contemporaries. The importance of his analysis was only recognized by the Chemistry Society with a Gold Medal some five years after they recognized Mendeleev.

Dimitri Mendeleev was a Russian Chemist and inventor who is credited with formulating the periodic law. Apparently, he was unaware of previous publications regarding the law of octaves of John Newlands. Mendeleev created his own periodic table of elements, which is much like the one we use today.

It is of interest to note that Gann frequently used terms and analogies in the *Ticker and Investment Digest* article related to chemistry and the elements. For example, we find the following passage:

"The power to determine the trend of the market is due to my knowledge of the characteristics of each individual stock and a certain grouping of different stocks under their proper rates of vibration. Stocks are like electrons, atoms and molecules, which hold persistently to their own individuality in response to the fundamental law of vibration. . . Science has laid down the principle that the properties of an element are a periodic function of its atomic weight."

In 1909, prior to the *Ticker and Investment Digest* magazine article, there was a series of advertisements posted in the *New York Herald* that made statements similar to those of W.D. Gann. The advertisements only provided a business name with the title "ORO-LO", along with a street address. Although there is no authorship attached to the advertisements, we know that they were made by Gann based on their content. In one advertisement dated Sunday April 18, 1908, the author states,

> "I have proved after nine years of scientific investigation that it is possible to know every move the markets make. It is a scientific problem, not guess work, as many believe. I have investigated all "Systems", found most of them worthless to the average trader and none of them perfect. I investigated astrology and kindred sciences to learn the law of the movements in the markets. In them all there was something lacking, and not until I struck upon the law of vibration and attraction as applied in Wireless Telegraphy did I find the key to Wall Street. I find the different stocks grouped into families, each having its own distinct vibration, which acts sympathetically upon others of the group and causes them to move in unison. I now have perfected my theory until I can forecast every move in Stocks, Cotton, and Wheat."

If I may summarize, we find Gann describing in both quoted passages, a method by which he groups stocks into what he describes as "families", which is based on their proper rates of vibration. In being consistent with Gann's time, I looked for definitions describing families of elements prior to 1909. I found that element families are comprised of a set of elements sharing common chemical properties, the members of the same family showing striking resemblances among one another.

One of the examples I found prior to 1909 described the chlorine family, which includes, besides chlorine itself, bromine, iodine, and fluorine. If you look at the Periodic Table below on the following page, you will see these elements in the second to last column, with atomic numbers of 17, 35, 53, & 9 respectively. To be perfectly honest with you, as I was researching these families, I expected to confirm Newlands Law of Octaves - that all of these elements would be a multiple of eight elements from each other - that the 8th element from Fluorine would be Chlorine, and the 8th from Chlorine would be Bromine, and so on. However, this was not the case.

If counting from Fluorine (9) to Chlorine (17), the theory holds up, but from Chlorine (17) to Bromine (35) it breaks down. The eighth from Chlorine (17) is Manganese (25), but it is not in the same family. Counting eight more elements from Manganese (25) is Arsenic (33). I wasn't expecting this at all. I had assumed that Newlands Law of Octaves and Mendeleev's Periodic Law were the same thing, but further research showed that this was not the case.

Newlands found that every eighth element had similar physical and chemical properties when they were arranged in increasing order of their relative masses. The arrangement of elements in Newlands' Octave resembled the musical notes. However, Newlands' Octaves could be valid up to calcium (20) only; as beyond calcium (20), elements do not obey the rules of Octaves. Thus, Newlands' Octaves were valid for lighter elements only. Despite the fact that Newlands Octaves breaks down as you progress through the periodic table, this theory is still at the foundation of music & light.

After the publication of the law of octaves by John Newlands in 1864, there are a few more publications worth mentioning

PERIODIC TABLE OF ELEMENTS

1	2	3	4	5	6	7	8	9	10	11	12	13	14	15	16	17	18
hydrogen 1 H 1.0079																	helium 2 He 4.0026
lithium 3 Li 6.941	beryllium 4 Be 9.0122											boron 5 B 10.811	carbon 6 C 12.011	nitrogen 7 N 14.007	oxygen 8 O 15.999	fluorine 9 F 18.998	neon 10 Ne 20.180
sodium 11 Na 22.990	magnesium 12 Mg 24.305											aluminium 13 Al 26.982	silicon 14 Si 28.086	phosphorus 15 P 30.974	sulfur 16 S 32.065	chlorine 17 Cl 35.453	argon 18 Ar 39.948
potassium 19 K 39.098	calcium 20 Ca 40.078	scandium 21 Sc 44.956	titanium 22 Ti 47.867	vanadium 23 V 50.942	chromium 24 Cr 51.996	manganese 25 Mn 54.938	iron 26 Fe 55.845	cobalt 27 Co 58.933	nickel 28 Ni 58.693	copper 29 Cu 63.546	zinc 30 Zn 65.39	gallium 31 Ga 69.723	germanium 32 Ge 72.61	arsenic 33 As 74.922	selenium 34 Se 78.96	bromine 35 Br 79.904	krypton 36 Kr 83.80
rubidium 37 Rb 85.468	strontium 38 Sr 87.62	yttrium 39 Y 88.906	zirconium 40 Zr 91.224	niobium 41 Nb 92.906	molybdenum 42 Mo 95.94	technetium 43 Tc [98]	ruthenium 44 Ru 101.07	rhodium 45 Rh 102.91	palladium 46 Pd 106.42	silver 47 Ag 107.87	cadmium 48 Cd 112.41	indium 49 In 114.82	tin 50 Sn 118.71	antimony 51 Sb 121.76	tellurium 52 Te 127.60	iodine 53 I 126.90	xenon 54 Xe 131.29
caesium 55 Cs 132.91	barium 56 Ba 137.33	57-70 *	lutetium 71 Lu 174.97	hafnium 72 Hf 178.49	tantalum 73 Ta 180.95	tungsten 74 W 183.84	rhenium 75 Re 186.21	osmium 76 Os 190.23	iridium 77 Ir 192.22	platinum 78 Pt 195.08	gold 79 Au 196.97	mercury 80 Hg 200.59	thallium 81 Tl 204.38	lead 82 Pb 207.2	bismuth 83 Bi 208.98	polonium 84 Po [209]	astatine 85 At [210]	radon 86 Rn [222]
francium 87 Fr [223]	radium 88 Ra [226]	89-102 **	lawrencium 103 Lr [262]	rutherfordium 104 Rf [261]	dubnium 105 Db [262]	seaborgium 106 Sg [266]	bohrium 107 Bh [264]	hassium 108 Hs [269]	meitnerium 109 Mt [268]	ununnilium 110 Uun [271]	unununium 111 Uuu [272]	ununbium 112 Uub [277]		ununquadium 114 Uuq [289]				

* Lanthanide series

lanthanum 57 La 138.91	cerium 58 Ce 140.12	praseodymium 59 Pr 140.91	neodymium 60 Nd 144.24	promethium 61 Pm [145]	samarium 62 Sm 150.36	europium 63 Eu 151.96	gadolinium 64 Gd 157.25	terbium 65 Tb 158.93	dysprosium 66 Dy 162.50	holmium 67 Ho 164.93	erbium 68 Er 167.26	thulium 69 Tm 168.93	ytterbium 70 Yb 173.04

** Actinide series

actinium 89 Ac [227]	thorium 90 Th 232.04	protactinium 91 Pa 231.04	uranium 92 U 238.03	neptunium 93 Np [237]	plutonium 94 Pu [244]	americium 95 Am [243]	curium 96 Cm [247]	berkelium 97 Bk [247]	californium 98 Cf [251]	einsteinium 99 Es [252]	fermium 100 Fm [257]	mendelevium 101 Md [258]	nobelium 102 No [259]

because they mention the special relationship between color & music. The first that I would like to mention is an article written by W. F. Barrett in the *Quarterly Journal of Science* dated January, 1870. There is a section in the article where he talks about the harmony of color and music. By comparing wave lengths of light with wavelengths of sound, not their actual lengths, but the ratio of one to the other, Barrett was able to show that they agree mathematically with respect to the septimal scale which divides them. To do this, he reduced the best determinations of color wave-lengths for his day into a common ratio, and compared the results with the wavelengths of the notes of the musical scale reduced to the same ratio. Below is a table reproduced from the book of the limits of wavelengths of the different colors of the spectrum as determined by a Prof. Listing.

TABLE OF WAVE-LENGTHS OF COLOURS IN THE SPECTRUM.
WAVE-LENGHTS: IN MILLIONTHS OF A MILLIMETER

Name.	Limit.	Mean.	Ratio.
Red	723 to 647	685	100
Orange	647 to 586	616	89
Yellow	586 to 535	560	81
Green	535 to 492	513	75
Blue	492 to 455	473	69
Indigo	455 to 424	439	64
Violet	424 to 397	410	60

Below is a table reproduced from the book showing the middle notes of the musical scale along with their wave-lengths and their reduction to a common ratio, taking the note "C" as 100.

TABLE OF WAVE-LENGTHS OF NOTES OF SCALE

Name.	Wave-length in inches.	Ratio
C	52	100
D	46 1/3	89
E	42	80
F	39	75
G	35	67
A	31	60
B	27 1/2	53
C_2	26	50

Putting the two ratios together, Barrett comes up with the following relationships:

RATIO OF WAVE-LENGTHS OF NOTES COMPARES TO RATIO OF WAVE-LENGTHS OF COLOURS.

Notes.	Ratio.	Colours.	Ratio.
C	100	Red	100
D	89	Orange	89
E	80	Yellow	81
F	75	Green	75
G	67	Blue and Indigo (mean)	67
A	60	Violet	60
B	53	[Ultra Violet	53]
C_2	50	[Obscure	50]

In regards to the table on the previous page, he has this to say,

> "Assuming the note C to correspond to the colour red, then
> we find D exactly corresponds to orange, E to yellow, and F
> to green. Blue and indigo, being difficult to localize, or even
> distinguish in the spectrum, they are put together: their mean
> exactly corresponds to the note G. Violet would then exactly
> correspond to the ratio given by the note A."

He goes on to point out that placing two colors nearly alike next
to each other is bad, just as it is well known that two adjacent
notes of the scale sounded together produce discord. Selecting and
sounding together two different notes may produce either discord
or harmony just as with the placement of certain colors next to
each other. As an example, he says that the notes D and E together
are bad, just as orange and yellow when contrasted. However, C
and G harmonize perfectly as do red and blue. Likewise, C and F is
an excellent interval, and so is the combination of red and green.

So what we appear to have is evidence of a sevenfold scale
in color and music. We could say that nature naturally divides
wholes into 7 distinct periods or phases. In keeping with this prem-
ise, William Fishbough writes the following in his book entitled,
*The End of the Ages: with Forecasts of the Approaching Political,
Social, and Religious Reconstruction of America and the World*,

> "Many years ago the present author composed and published
> a volume [The Macrocosm and Microcosm] in which an at-
> tempt was made to show, that the number of degrees in each
> and every complete scale of evolution, is seven: that the order
> of their sequence is the same as the order of the seven notes
> of the diatonic scale in music, and the seven colors of the rain-

bow, with their harmonics and complementary relations; and that the whole system of creation, constructed on this plan, presents a grand series of octaves any one of which, being ascertained, would, in a general way, serve as a type and exponent of all the others, whether upon a higher or lower scale."

Fishbough continues by referencing Prof. Barrett's work of which I have already made mention. On page 10 of Fishbough's book, he writes the following,

"After pointing out the correspondences in the series and progressions in the two scales, Prof. Barrett adds, in a foot note, this striking remark:

""This," says he, "appears to be a fundamental law of the universe, viz: That an original impulse of any kind finally resolves itself into periodic motion.""

Compare this to what Gann states in his *Ticker and Investment Digest* article from 1909,

"Science teaches us that an original impulse of any kind finally resolves itself into periodic or rhythmical motion . . ."

In Fishbough's first example of testing his hypothesis, that history proceeds in regular cycles in which, from first to last, there is a sevenfold series of differential parts or stages exactly answering to the seven distinctive degrees in the music and color scales, he explains that he could not seem to find anything that supported his theory until one day when he was looking over an old table of chronology of the American Republic. He saw what appeared to be something like a regular succession of waves or steps, so to speak, in the development of our own national history. More consideration revealed the fact these waves or steps ran in periods of 12 years. He then proceeds to describe these 12 year periods, in which 7 of these would complete an octave of 84 years.

To each of these 7 periods Fishbough assigns a certain set of characteristics that are unique to each stage, and then commencing with the year 1776, describes each 12 year block in relationship to the set of characteristics assigned to each. These periods are listed below.

1st: The Revolutionary and Chaotic Period
2nd: The Organizing Period
3rd: The Testing Period
4th: The Median Period
5th: The Period of Ideas and Aspirations
6th: The Period of Fruitage
7th: The Period of Ripeness

Starting on page 16 of his book, he describes each of these periods with the aid of historical events that took place during each with respect to the United States. It is implied that in 1776 when our national independence was declared, this was the original impulse that set in motion the periodic events to shape this nation.

It is worth mentioning that the end of the 7th period of 12 years is 1860, which would begin a new set of periods commencing with the Revolutionary and Chaotic. With this in mind, around 1858 to 1859, Fishbough ventured a prediction based on his research that the year 1860 would witness a change in our nation which would in some sense answer to a national death. It is well known that the presidential election of 1860 set the stage for the

American Civil War. The nation was divided on the issue of slavery, and chaos & revolution reigned. These are the very characteristics that he used to describe the events corresponding to the first period of any cycle.

If you add an additional 84-year period to 1860, we come to 1944. It marks the end of the 7th period in another 84-year cycle, which would run from 1932 to 1944. Interestingly, this time period is also identified by Gann in his novel, *The Tunnel Thru the Air*, where he writes on page 83, "Another bad period for the United States will be 1940 to 1944." If you add another 84-year period to 1944, we come to 2028. In my first book, I wrote 2016, but attentive readers brought this to my attention as this was a miscalculation and error on my part.

In keeping with the hypothesis outlined by Fishbough, we find another example of this septimal division in the work of Dr. Jos. Rhodes Buchanan. In his book entitled, *Periodicity: The Absolute Law of the Universe*, he tells us that after retirement from a Medical College in 1856, he was attracted by an apparent periodicity in nature in the phenomena of disease and in the different influences of week days, months and years, and even in his own affairs in the college. He further states,

> "Popular opinion fixed upon the sixth day of the week, Friday, as unlucky, and some of my experiments seemed to sustain that idea, which was expressed in the creative legend of Genesis, that God was fatigued on the sixth day and rested on the seventh, which was therefore ordered to be a day of rest.
>
> "Friday, the sixth day, was the day of the crucifixion of Jesus, and has since been regarded as hang-man's day, and used for that purpose. The wide spread opinion that Friday is an inauspicious day, would not have been so long maintained

without some foundation in nature, and the same impression as to the number thirteen must have been based on some experience."

From these observations he had worked up his theory on periodicity and put it to the test. He goes on to say,

"To make decisive tests of the law, I have been accustomed upon first meeting a stranger to tell him of the favorable and unfavorable periods of his life, and to find him astonished at the revelation of his troubles, the times of deadly sickness, financial loss, disappointments, calamities and failures in schemes that looked plausible."

Then, he goes on to say,

"The law which I have found in operation, and which my most intimate friends, in testing, have become convinced by experience that it is a law of great importance to be understood, is easily stated. It is this - that all vital operations proceed in varying course, measured by the number seven. This septimal division I expect to find in the life of every individual from youth to age, in the progress of diseases, in the history of nations, societies, enterprises, and everything that has progress and decline - in short in all life, for all life has its periods of birth growth decline and death."

Buchanan goes on to explain that an individual's life is governed by 7 periods of 7 years each, which amounts to 49 years in total. After the 49th year, the cycle repeats itself. Each of these 7 periods is given the name of a day of the week for easy identification. Thus, Sunday is the 1st period, Monday the 2nd, and so on to Saturday, which corresponds to the 7th.

In addition to the 7-year periods, each year within the 7-year period also corresponds to a day of the week. Thus, the first year is a Sunday year, the second year a Monday year, etc., so that like Gann says, there is a wheel within a wheel. A person could be in a Sunday 7-year period, but in a Tuesday year. Then, the year is also broken down into 7 periods of approximately 52 days each. So this would be a wheel within a wheel within a wheel. Earlier, I provided you with the passage where Buchanan explains that the sixth day was considered evil or unfavorable because it was the day that Jesus was crucified. So the start of the 6th period is a time to look out for things of an undesirable nature to occur.

In addition to what has been stated thus far, Buchanan provides another unique example behind the theory of the 6th period being unfavorable. Dividing the year into 7 periods of approximately 52 days each, we find that the 6th period would begin on approximately the 260th day and end on day 312. When you consider that it takes 9 months or 40 weeks from conception to birth, which is approximately anywhere from 267 to 280 days, it falls right in the middle of the 6th period from conception. As Buchanan describes it in his book,

> "These rules show that in serious diseases the crises arrives on the 6th, 13th and 20th days - first on the 6th to the 8th day, the moon passing through one-fourth of its orbit - 2nd on the 13th to the 15th day, as she passes through half of her orbit - and third, the 20th to 22nd day, the moon passing through the end of its third quarter, having passed through 270 degrees.

"This illustrates the periodic law first stated in this book, discovered over thirty years ago - the fateful six in the number seven - and the fateful 270 - the number of days which brings us to separation from our mother and exposure to a period of danger."

Throughout the book, we find that the major theme is that the 6th period is the most evil. However, from the middle of the 4th period to the end of the 7th is also described as unfavorable because it corresponds to the latter half or decline of the year or day if starting the cycle at the spring equinox or sunrise.

Like Fishbough, Buchanan also provides us with examples of the application of these periods starting when the United States began as a nation on July 4th, 1776. In the cycle of 49 years, the 6th period would be from 1811 to 1818 and it is within this range Buchanan says that the unsatisfactory war with England called the war of 1812 came. Interestingly, another 49 year period would give us the 6th period in the next cycle from 1860 to 1867. I have already made mention of this time frame as it pertains to the presidential elections of 1860 that divided the nation over slavery and was the catalyst for the Civil War. The next 49-year period gives us the 6th period in the next cycle from 1909 to 1916 corresponding to U.S. involvement in the World War.

Buchanan also applies his rules of periodicity to the life of Napoleon Bonaparte. He provides his date of birth as August 15, 1759. He breaks up his life into good and evil periods based on the year of his birth. The good periods are Sunday through the first half of Wednesday, and the evil are the latter half of Wednesday through Saturday periods. I will not go into the specifics of the yearly analysis, for the main point I would like to draw out is related to the division of the year of 365 days into 7 periods of a proximately 52 days each. Buchanan identifies May and June as his evil months and a quick perusal of the table below will show that the 6th Period of the Year runs from May 2 to June 23.

1st Period	Sun	Aug 15 - Oct 6
2nd Period	Mon	Oct 6 - Nov 27
3rd Period	Tue	Nov 27 - Jan 18
4th Period	Wed	Jan 18 - Mar 11
5th Period	Thu	Mar 11 - May 2
6th Period	**Fri**	**May 2 - June 23**
7th Period	Sat	June 23 - Aug 15

It is of interest to note that this same procedure of dividing the year into 7 periods starting from the day and month of birth appears to be the way Gann identified periods for Wilhelm Hohenzollern in the 1919 article.

First and foremost, I would like to give thanks to the Gann Study Group for the publication of the e-book entitled, *W.D. Gann on the Law of Vibration*. I neglected to do so in my first publication, but it was in this e-book that I first read the 1919 article. Needless to say, after reading the article I was in awe. This article is one of the catalysts for the direction I have taken with my research into Gann. In this 1919 article, Gann makes some predictions about the German Kaiser, Wilhelm Hohenzollern, and provides some details as to how he made his predictions. In one of the passages Gann lists a series of dates indicating that they would be Wilhelm's most evil periods for the year in question. Gann writes,

"The following are his most evil periods for this year:- March 20 to 27, May 10 to 14, July 2 to 5, August 23 to 25, October 10 to 13 and November 7 to 13."

One of the things I had found is that the dates were separated by approximately 52 days. It appeared as if Gann had taken the year of 365.24 days and divided it into 7 equal parts of 52.177 days each, and starting from Wilhelm's birth month and day, which is January 27, added 52.177 day periods to each of the resultant dates. Doing so would give you the following dates on the left.

Microsoft Excel	Dates in the 1919 Article
January 27	
March 20	March 20 to 27
May 11	May 10 to 14
July 2	July 2 to 5
August 23	August 23 to 25
October 14	October 11 to 13
Dec 6	November 7 to 13

Compare the dates calculated in Microsoft Excel with the dates from Gann's article on the right. Note that Gann doesn't include the first period in his list of dates from the article. That said, the only major difference between the two lists is the last date range given. Gann lists dates of November 7 to 13 as opposed to Dec 6. This date range in November just so happens to be in the middle of the 6th period corresponding to the November 9th date when Wilhelm abdicated. Regarding this, the article states,

> "Observe that he abdicated on his evil day, the ninth, in his evil month, November."

Here we see that Gann's procedure is the same as Buchanan's, and he even uses the same terminology.

In addition to what I found above, there is a passage in Buchanan's book that closely matches a passage in the 1919 Gann article. In the article, Gann opens the analysis by saying,

> "Wilhelm Hohenzollern, the famous imperial scoundrel, whose crimes against women and children have debauched and shocked the civilized world and caused him to be the most hated and despised man in history, was born January 27, 1859."

Now consider what Buchanan writes on page 124 of his book concerning Napoleon Bonaparte.

> "But I must select one famous example, in the life of that imperial scoundrel, Napoleon Bonaparte, whose crimes have debauched the world's conscience so completely that he still receives a tribute of admiration."

This, written in 1897, appears to be modeled by Gann to use in the opening of his analysis of the German Kaiser in the 1919 article. The statements are very similar. We can deduce that Gann most likely read and studied Buchanan's book at some point, and there is more evidence to support this claim.

On page 132 of *Periodicity*, in regards to his analysis of Napoleon Bonaparte, Buchanan writes,

> "Blind to his real condition he rose again in March 1815, widely detested and met his fate at Waterloo, abdicating in his fatal month June, surrendering to England in July and imprisoned at St. Helena, dying in his evil month May 5, 1821, going to a world not entirely congenial to his nature."

In the 1919 article, Gann makes a statement that appears to be in reference to this passage by Buchanan when he writes,

> "Had the former Kaiser understood the science of letters and numbers he would have realized that he would meet his Waterloo through Woodrow Wilson, whose name stands for justice and liberty."

This passage appears to be in reference to Buchanan's description of how Napoleon met his fate at Waterloo. With all that has been presented, it certainly appears that Gann used Buchanan's method in some part with respect to the 1919 article.

Now, what if I told you that I was able to find another example where it appears that Gann used this same method to derive key periods? This is exactly what I found when analyzing the periods mentioned in Gann's book entitled, *Face Facts America! Looking Ahead to 1950*. At the end of the book on page 44 of the version I have in my possession, Gann dates the book, May 24, 1940. On page 9 he even refers to this date saying, "As I write on May 24, 1940 . . ." On page 30, he has a section with the heading entitled, "When Will the War End?" This is the portion of the text where we will find our next example. In this section, Gann writes,

> "By the use of these Master Time cycles, which are based on the law of mathematics and a repetition of time cycles that I have discovered, I can forecast important culminating periods in the history of countries or wars just the same as I forecast important tops and bottoms in the stock market. Using the dates of the beginning of the World War in 1914 [July 28], and the end of the World War on November 11, 1918, which I predicted by using these cycles, then from the beginning of

the present war on September 1, 1939, when Hitler invaded Poland, I get the following important periods on the present war:

"May, 1940 - On May 19 in the New York Herald-Tribune and the New York Journal-American, I advertised that the tide would turn against Germany on May 25 and that they would meet with greater reverses by August 10th, 1940.

"July and August, 1940 - A very unfavorable cycle runs against Germany from July 10 to August 10 and the war should go against Germany at that time. There is a possibility that peace could come in August, as the time cycle indicates a possibility of the end of the war by September 1, or after it has run one year."

Before we get into analyzing these dates, I would like the reader to realize that it is not so important whether these predictions were accurate, but what is most important is to understand why and how he came up with these dates in the first place. It is the method that we are looking for.

Now, Hitler invades Poland on September 1, 1939 to begin the War. If we take the solar year of 365.24219 days and divide it into 7 periods, each period will be 52.177 days in length. Starting with the date for the beginning of the war and using a generic time of 12:00 noon, we get the following periods:

1st Period	Sun	Sep 1 - Oct 23
2nd Period	Mon	Oct 23 - Dec 14
3rd Period	Tue	Dec 14 - Feb 5
4th Period	Wed	Feb 5 - Mar 28

5th Period	Thu	Mar 28 - May 19
6th Period	**Fri**	**May 19 - July 10**
7th Period	Sat	July 10 - Sep 1

Since the original impulse was Hitler's invasion of Poland, we see that the evil 6th period in relationship to this start date begins on May 19, 1940. Referring back to the quoted passage from Gann's book, we find that it is on May 19 where Gann says he advertised that the tide would turn against Germany on May 25, which is just within the start of the evil 6th period. Furthermore, in the next passage Gann says that "A very unfavorable cycle runs against Germany from July 10 to August 10". July 10, 1940 is the exact start date of the 7th period when measured from the start of the War. He also says there is a possibility peace could come in August. This is consistent with the characteristics of the 7th period, which is symbolic of the sabbath, which is a time of retreat or rest. In one of Gann's courses in a section on the human body, Gann writes,

> "There are seven openings in the head - two eyes, two ears and two nostrils, equally divided, three on each side. From this we get our Law of Three and know the reason why the change comes after two and in the third period. The seventh opening in the head is the mouth and everything goes down. Study your seven-year periods and see how your markets go down and make tops and bottoms."

In this same course he also writes,

> "Then you will understand why the children of Israel marched 7 times around the walls of Jericho, blew the ram's horn 7

times and the walls fell down on the 7th day. This Law is also backed with astrological proof, but anything that can be proved in any way or by any science is not correct unless it can be proved by numbers and geometry."

In Gann's discussion of United States Steel from one of his courses, Gann says,

"Going into the 7th year indicated lower prices. The 7th year is always a year for a panicky decline . . ."

This tells us that in addition to the 6th period being evil, the 7th period seems to always be associated with peace, rest, reversals, and declines.

REVISITING THE PERIODIC LAW

Earlier, I had quoted Gann from the *Ticker and Investment Digest* article where he said,

"Since all great swings or movements of the markets are cyclic they act in accordance with periodic law."

I had also mentioned that I thought this statement to be very telling because in addition to saying that movements in the markets are cyclic, he is also saying that they act in accordance with periodic law. As previously stated, I had assumed that Newlands Law of Octaves and Mendeleev's Periodic Law were the same thing, but further research showed that this was not the case. Newlands found that every eighth element had similar physical and chemical properties when they were arranged in increasing order of their relative masses. The arrangement of elements in Newlands' Octave resembled the musical notes. However, Newlands' Octaves could be valid up to calcium (20) only; as beyond calcium (20), elements do not obey the rules of Octaves. Thus, Newlands' Octaves were valid for lighter elements only. The Periodic Law encompasses more than just Newlands' Octaves.

We can deduce that Gann got much of his inspiration concerning the Periodic Law from a book entitled, *The New Knowledge* by Robert Kennedy Duncan. In the opening pages of this book, we find the dedication which is very similar to the dedication that Gann uses in his novel, *The Tunnel Thru the Air*. This first Screen shot is from Duncan's book, *The New Knowledge.*.

TO THE MEMORY OF MY MOTHER

This second screen shot is from Gann's book, *The Tunnel Thru the Air*.

DEDICATED

TO THE MEMORY OF MY MOTHER

SUSAN R. GANN

AND

TO AN OLD SCHOOLMATE IN TEXAS

MY NATIVE STATE

In addition, we can also find on pages 22 to 23 of this book direct passages that Gann uses in the *Ticker and Investment Digest* article.

"Briefly and technically, the law states that *"the properties of an element are a periodic function of its atomic weight."* This is a very concise statement indeed of an extraordinary fact. The statement means no more nor less than this: That if you know the weight of the atom of the element you may know, if you like, its properties, for they are fixed. Just as the pendulum returns in its swing, just as the moon returns in its orbit, just as the advancing year ever brings the rose of spring, so do the properties of the elements periodically recur as the weights of the atoms rise."

If we now consider that Newlands' Octaves was valid for lighter elements only, then what about the heavier elements beyond calcium (20)? A mathematical formula or sequence if you will, explains the whole arrangement of the elements including both the lighter and heavier elements. The formula is $2(n^2)$ where n is an integer greater than zero.

When n is equal to 1, the result is 2. Referring back to the Periodic Table on page 14, the lightest element is Hydrogen (1). If you count two spaces from Hydrogen you will get to Lithium (3), which is in the same column, but is not in the same family. When n is equal to 2, the result is 8, because 2^2 is 4, and 4 X 2 = 8. On the Periodic Table, starting with Lithium (3), count eight spaces over and you will get to Sodium (11) in the next row. Thus, Lithium (3) and Sodium (11) are in the same family and have similar properties. Pay very close attention to what is stated in *The New Knowledge*. On page 22 it says,

". . . the number of elements between any one and the next similar one is seven. In other words, members of the same groups stand to one another in the same relation as the extremities of one or more octaves in music!"

The number of elements between any one and the next similar one is seven. So if our any one and the next similar one is Lithium (3) and Sodium (11) we find that there are seven elements in between them. **It is the elements between two elements of similar properties that are being likened to an Octave in music**. What was being conveyed from that passage is shown below.

Elements with Similar Properties	Elements in Between	Notes
Lithium (3) -	- - -	-
	Beryllium (4)	C
	Boron (5)	D
	Carbon (6)	E
	Nitrogen (7)	F
	Oxygen (8)	G
	Fluorine (9)	A
	Neon (10)	B
Sodium (11) -	- - -	-

Consider that there are seven notes in the musical scale as listed above. If we start at C, then the note above C is also named C. This is an octave. It is natural to think that if we thought of Lithium (3) as comparable to the note C, then Sodium (11), which has similar properties, would correspond to C above. Yet, if we label each of the elements after Lithium (3) with their corresponding note, we would be one short. The next C would correspond to Neon (10). It is obvious that in the seven note musical scale, there are six notes between C below and C above, but in the Periodic Table there are seven. From all of this we learn that nature naturally divides wholes into eight distinct parts, with the ninth being similar in properties and characteristics to the 1st. We can see this in the list of elements from Lithium (3) to Sodium (11) and from Sodium (11) to Potassium (19) from the Periodic Table on page 14.

1st	Lithium (3)	Sodium (11)
2nd	Beryllium (4)	Magnesium (12)
3rd	Boron (5)	Aluminum (13)
4th	Carbon (6)	Silicon (14)
5th	Nitrogen (7)	Phosphorus (15)
6th	Oxygen (8)	Sulfur (16)
7th	Fluorine (9)	Chlorine (17)
8th	Neon (10)	Argon (18)
9th	Sodium (11)	Potassium (19)

Keep in mind that eight is not the only number that nature uses to divide wholes. As I mentioned before, the formula being utilized is $2(n^2)$ where n is an integer greater than zero. So when n is equal to 3, the result is 18, and if we continue in like manner, we get a mathematical sequence of the following numbers where n is equal to 1 through 9, (2, 8, 18, 32, 50, 72, 98, 128, 162).

On the Periodic Table, we find that the maximum number of elements between one and one with similar properties is 32. In addition, we find this mathematical structure not only in the increasing order of atomic weight of the elements, but also in their geometrical structure. This can be seen in the rules governing electron shells.

In chemistry and atomic physics, an electron shell, also called a principal energy level may be thought of as an orbit followed by electrons around an atom's nucleus. The closest shell to the nucleus is called the "1 shell", followed by the "2 shell", and so on farther and farther from the nucleus. Each shell can contain only a fixed number of electrons. The first shell can hold up to two electrons, the 2nd shell can hold up to eight, the 3rd shell can hold up to 18, and so on. This is representative of the formula $2(n^2)$.

What I want the reader to take away from this is that if the number of elements between elements of similar properties is seven, and if seven is an important number, then why wouldn't the other numbers that separate elements of similar properties be just as important. These numbers would follow the mathematical sequence $2(n^2) - 1$. Where n is an integer greater than zero, the resultant sequence would be 1, 7, 17, 31, 49, etc. Would not all of these numbers hold some special significance along with the number seven? If not, then maybe it was not the number seven that Gann held in special esteem as a result of the Periodic Law, but maybe it was something else entirely.

Consider if you will, that instead of the theory that nature naturally divides wholes into 7 distinct parts as I mentioned in my previous publication, and also on page 17 of this book - that instead, it naturally divides wholes into 2, 8, 18, 32, or even 50 distinct parts. This is especially interesting with respect to Gann's division of the circle, price ranges, and time periods into eighths. Consider that he often said that the halfway point was the most important division, which is a division of a whole into two parts. In eighths, this would correspond to 4/8ths. It is also known that he divided price ranges and time periods into thirds, but I would like to continue with the division of eighths because of an interesting relationship I found to the Law of Vibration.

EIGHT PARTS AND THE LAW OF VIBRATION

One of the more interesting things that I found during my research are the comments of an individual named J. H. Schucht. In *English Mechanic and World of Science:* No. 1320, dated July 11, 1890, there is a section on page 425 that reads, THE PHYSICAL BASIS OF MUSIC. In this section, J. H. Schucht writes,

> "Readers of "Ours" may perhaps recollect that I have for 20 years repeatedly stated that there is only one law which governs all movements and vibration in nature - *i.e.*, harmony, and that acoustics is the child born of it. I have not shrunk from laying myself open to be called a paradoxer by saying that our earth, Solar system, and the whole universe follows this simple law. Finding that acoustics is only an outcome of this law, I have called it the law of vibration."

Schucht goes on to describe his experiment of setting a rod into vibration and noting that whether connected to other bodies or set apart from them, it vibrates in the same relative proportions. In relationship to the above, In *English Mechanic and World of Science:* No. 1711, dated January 7, 1898, J. H. Schucht writes,

> "In 1873 I made an apparatus by which I illustrated the law which I call the law of vibration. I also made a globe on which I marked those proportions which might possibly result from the law of vibration. . . Since that time I have occasionally written letters on the subject in "Ours," in the hope that some of my fellow readers with more time and ability than I can command would take the matter up. In those letters I stated that all bodies of matter in nature vibrate in eight equal parts

or volumes if the body be homogeneous. If not homogeneous, those parts furthest away from the centre of gravity of the body are, as a matter of course, less in volume than those nearer it. . . As periodicity can only take place in nature with movement of matter, it must follow this law of vibration. . . When Kepler talked about "Spheren Musik," he had this law in his mind. These periods are not caused by the moon's rotation; she only follows the same law in company with other matter in nature."

What is notable is the fact that he states that all bodies of matter in nature vibrate in eight equal parts if the body be homogeneous. Furthermore, from what I can understand of his experiments, he is saying that Periodicity or cyclic phenomena is a result of this law, not the cause of the Law. If we look to the cycle of the moon to explain the cause of certain Periodic phenomena in nature, we are in error. The moon's periodicity is caused by the same law that controls other periodicities in nature. Likewise, we have to consider that the same can be said of other planetary cycles.

In summary, maybe these eight parts have distinct characteristics like each of a series of elements on the Periodic Table. For example, from Sodium (11) to Argon (18). Maybe these divisions react with each other in specific ways just like the elements of the Periodic Table react with each other. Maybe there is more than meets the eye with respect to the division of a whole into eight parts. This reminds me of my early researches into Astrology and the original division of the heavens into eight houses. It was in Cyril Fagan's book entitled, *Astrological Origins* where I learned of the Greek Oktotopos, which means eight places.

THE OKTOTOPOS (EIGHT PLACES)

On page 164 of *Astrological Origins* , Fagan writes,

"In the original scheme of things, as conceived by the early Egyptians, these so-called houses or places were not measures of space at all but *measures of time*; a fact which modern astrological mathematicians have utterly failed to grasp."

It is from the activities of the cycle of a day as a measure of time that the houses or places get their meanings and characteristics. Many years ago I started work on a manuscript to explain from a unique perspective how the astrological houses got their meanings and characteristics, but never published the material. It was also based on the division of time into eight equal parts. In relationship to this, there are other facts concerning the division of a whole into eight parts that will prove to be very enlightening. This comes from Egyptian Cosmogony.

COSMOGONY

A cosmogony can be defined as a model of the origin and evolution of the universe. It explains how the objects of the physical world come into manifestation. Since it traces the evolution of these objects from the beginning stage of creation, it also serves as a map of the underlying structure and order of the universe. In ancient Egypt, the model of the origin and evolution of the universe can be found in the material concerning the Paut Neteru.

PAUT NETERU

In ancient Egypt, the word paut was used to describe the essence of a thing. Therefore, in relationship to the creative process, it refers to unformed energy/matter. Neter, the root of the word Neteru, refers to the Supreme Being. Neteru, the plural form of Neter, refers to the manifestations or expressions of the Supreme Being. Therefore, the Paut Neteru conveys the idea of a model describing the creative process from its unformed state into its various expressions.

These expressions are derived from the fact that creation is a process of differentiation. Within this process there are two extremes. The first is a state in which no things exist. It is the beginning of the creative process. The second is a state in which things do exist. It is the end result of the creative process. The things that take place between these two extremes are controlled by a set of faculties recognized as the attributes of the Supreme Being. Thus, as creation unfolds, the Supreme Being differentiates energy into its many attributes, each possessing a unique role within the process. As you will see, each role is centered on the maintenance of order in the universe.

The order that we perceive in the universe is necessary to maintain harmony between the various objects of its system. Therefore, the structure of the universe is based on a concept known as interdependence. It is based on the fact that certain forms are created to perform specific functions. Thus, the universe is composed of different objects working together and depending on each other to carry out a similar goal.

Since creation is a process of differentiation, the objects that exist at the end of the creative process are nothing more than modifications of the original state of energy. Take a look at the diagram below. The circle at the top represents the state where no things exist. At the bottom is a triangle, rectangle, and square, representing the objects of the physical world at the end of the creative process.

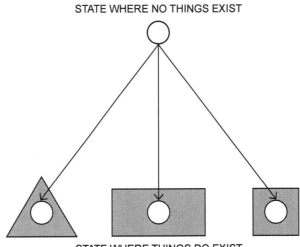

The different shapes show how the objects of the physical world vary in their form. The small circles inside each shape show that they contain the original Consciousness from which they are derived. Knowing that the objects of the physical world are modifications of the original state of energy, the Egyptians developed their cosmogony with the use of symbols representing the various attributes of the Supreme Being. Although several cosmogonies existed throughout the history of Egypt, there is one collection that stands out among the rest. The diagram on the following page is a picture of the Paut Neter or Tree of Life.

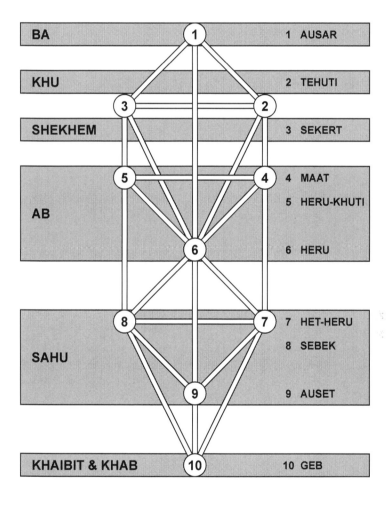

0 AMEN

BA   1 AUSAR

KHU   2 TEHUTI

SHEKHEM   3 SEKERT

4 MAAT

5 HERU-KHUTI

AB   6 HERU

7 HET-HERU

8 SEBEK

SAHU   9 AUSET

KHAIBIT & KHAB   10 GEB

Each Neteru is responsible for a role in the creative process as labeled through the spheres 1 - 9. The sphere labeled 0 is the state where no things exist, and the sphere labeled 10 represents the end result of the creative process. As it is beyond the scope of this section to go into a detailed description of the metaphysical properties of each sphere, it will be sufficient to concentrate on the function of each faculty as it relates to its role in the creative process.

Amen

Amen corresponds to the sphere labeled "0" on the Tree of Life. As stated, it represents the beginning stage of the creative process where no things exist. Since differentiation has yet to take place, there are no distinct objects to perceive. This concept is expressed in the philosophy of many cultures, and it is no different in Chinese cosmology. In *The Numerology of the I Ching*, Huang writes,

> "According to ancient Chinese cosmology, before creation there was nothing, the void".

The word "amen" became a fitting metaphor to describe this aspect of the Supreme Being because in everyday language the word was used to denote that which is "hidden" or "concealed". If an object is hidden or concealed, it is imperceptible. In harmony with this concept is another metaphor that the Egyptians used to describe this state. It is called "Tem", and literally means negative being or negative existence. It fittingly reinforces the concept that there is nothing in this state to perceive.

Ausar

Ausar corresponds to the sphere labeled "1" on the Tree of Life. The literature tells us that in the form of Ausar the Supreme Being takes on the role of creator. Thus, in one of the versions of the creation story from E.A. Wallis Budge's book entitled, *The Gods of the Egyptians Vol. 1: Studies in Egyptian Mythology*, we find on page 300, Neter or Neb er Tcher (Lord of the World) saying,

> "I was the creator of what came into being, that is to say, I formed myself out of the primeval matter, and I formed myself in the primeval matter. My name is Ausares [Ausar], who is the primeval matter of primeval matter".

In the above quoted passage, when it says that Ausar forms itself out of primeval matter, the "first" matter is of course the state where no things exist, Amen. Thus, in the first act of manifestation, Neter establishes itself in the role of creator. Ausar is Consciousness, which has the role of willing energy/matter to differentiate.

The grand metaphor that corresponds to this sphere is unity. It reinforces the fact that differentiation has yet to take place. In harmony with this concept is the hieroglyphic representation of the name Ausar, which contains a picture of a single eye. Among other things, it is telling us that Ausar is omnipresent.

This concept was depicted as the small white circles positioned within each of the geometric objects at the end of the creative process. If Ausar is omnipresent, then symbolically it can see all of the things that you do. In addition, since Ausar/Consciousness resides in all things, it is the common source of unity between all that is created. Ausar is the unifying principal in the universe.

In the next stage of manifestation, the creator Ausar brings forth its creative faculties as an act of differentiation. Therefore, in the creation story, Ausar is made to say the following on page 317 of Budge's *The Gods of the Egyptians Vol 1.*,

> "I, even I, spat in the form of Shu, and I emitted Tefnut, and I became from god one gods three, that is to say, from myself two gods came into being on earth this".

Ausar/Consciousness wills unformed matter to divide into the first two attributes. In the above passage they are identified as Shu and Tefnut.

The word "Shu" is etymologically related to a number of Egyptian words that denote fire, heat, light, and dryness. The word "Tefnut" is etymologically related to a number of Egyptian words that denote moisture. As symbols, they are synonymous with the Chinese concept of Yin and Yang, and represent another similarity in both Chinese and Egyptian cosmology. Also on page 2 of *The Numerology of the I Ching*, Huang writes,

> "According to ancient Chinese cosmology, before creation there was nothing, the void. That state of nothing was called Tai Chi . . . Eventually Tai Chi differentiated into two primary energies, the yin and the yang".

Shu represents the Yang, dry, masculine energy, and Tefnut, the Yin, moist, feminine energy. As creative faculties, they represent Ausar's means for creating the universe. On the Tree of Life, Shu and Tefnut correspond to the second and third spheres, which correspond to Tehuti and Sekert respectively.

Tehuti

Since the Egyptian Sages developed a cosmogony by referring their intuitions of the creative process to the mental and psychical processes in Man. They would have no doubt extended this to the physical processes, as the Sages likened the creation of the universe to the creative process between a man and a woman in their efforts to create a child.

In order to create a child, we know that it takes the interaction of two fundamental forms. One is the male, yang creative form, and the other is the female, yin creative form. Since Tehuti corresponds to masculine energy in its relationship to Shu, the second sphere corresponds to the male creative faculty and is therefore referred to as the "Will of God".

The will is nothing more than the ability to choose or make a decision because it is a potential act. Therefore, this sphere corresponds to a choice or potential act of God. This is important. The things that Ausar decides or chooses to create are always in harmony with its function, the maintenance of unity. This is why Neter in the form of Tehuti is omniscience or all knowing.

As the "God of Wisdom", the counsel given by Tehuti is always the best possible advice in regards to a situation. This is why Tehuti is patron of the Oracle, a device used to consult the Will of God regarding a potential action. If your action is in harmony with the Will of God, then your act will create unity, peace, and harmony. As stated, the things that Ausar decides or chooses to create are always in harmony with its function, the maintenance of unity.

Sekert

As stated, it takes the interaction of both a male and female to create a child. While the second sphere corresponds to what Neter wills to create, the third sphere corresponds to the power to actually carry it out, for it is the female who gestates and gives birth to the child (what was willed). That is why Neter in the form of Sekert is said to be omnipotent or all-powerful.

The Egyptian Sages understood that the ability to conceive a child is subject to a time constraint dictated by the female. This time constraint takes on the form of the female menstruation cycle. It averages approximately 29.5 days and is marked by two key points. They are the times of ovulation and menstruation.

The only time that a female can conceive is during ovulation, which occurs at the midpoint between menstruations. In order to plant the seed at the right time to ensure conception, the cycle requires you to plan. Interestingly, in Budge's hieroglyphic dictionary, the word "skher (sekher)" means, "to plan". It is etymologically related to Sekert.

After Ausar brings forth its creative faculties, it looks towards creating the objects of the physical world. This is achieved in six acts corresponding to spheres 4 through 9 on the Tree of Life. Just as a man and woman must interact to create a child, Tehuti and Sekert also interact to initiate this process. Likewise, in Chinese cosmology, we find on page 2 in *The Numerology of the I Ching*,

> "After two primary energies were generated, yin energy and yang energy interacted".

In the Egyptian literature, this interaction takes place when it is said that Tehuti speaks the hekau (words or power) residing in the third sphere. Symbolically, the hekau are the eggs that contain the genetic makeup of the entire universe, and the action of speaking (sound vibration) is said to fertilize them.

Maat

Maat corresponds to the sphere labeled "4" on the Tree of Life. Earlier, I mentioned that the objects of the world exist in their respective forms in order to perform a necessary function. For example, a lion has claws and teeth that are made to cut and tear through raw flesh. Combine this with the distinct shape and length of its digestive tract, and you have an animal whose form allows it to perform the function of being a carnivore.

Form and function are responsible for the interdependence that we see at work in the universe. This is the fundamental basis of Divine Law to which Maat corresponds. It is here that Neter creates the Divine Laws that the physical world will be bound to.

These Divine Laws help to sustain the proper functioning of the entire universe. Thus, before creation of the objects of the physical world, the creator has to first establish the laws that they will be bound to. Therefore, in the creation story, on page 309 in *The Gods of the Egyptians Vol. 1*, the creator is made to say,

"I laid a foundation in Maa [and] I made attribute every".

Since the 4th sphere is directly born out of the third, these laws are related to the various cycles found in nature.

Heru-Khuti

Heru-Khuti corresponds to the sphere labeled "5" on the Tree of Life. In this act, the creator takes on the form of Heru-Khuti to enforce the Divine Laws created in the preceding stage. What good is there for a law if there is no means of enforcing it? This ensures that if the Divine Laws are broken, there is something in place to restore the Divine Order.

In addition to enforcing the Law, Heru-Khuti also protects those who uphold it. This is why Heru-Khuti is often depicted carrying weapons such as knives, spears, and chains. Heru-Khuti is the patron of warriors, and its role is similar to the one that the police and military play in our current society. They also make use of various weapons to restore order, and also protect those who uphold the law. Thus, the grand metaphor for this sphere is "Divine Justice".

Heru

Heru corresponds to the sphere labeled "6" on the Tree of Life. In this act, the creator brings forth its main faculty for being able to function in the world. The Egyptians understood that Neter creates the world so that it could live and express itself through it. Thus on page 309 in *The Gods of the Egyptians Vol. 1*, we find a passage in the creation story where the creator says,

> "I brought [into] my mouth my own name, that is to say, a word of power, and I, even I, came into being in the form of things which came into being".

It is telling us that Neter comes into being in the form of the things that it creates, including us. As stated before, within the objects of the physical world there exists the original state of Consciousness as depicted previously.

Neter would not be omnipotent if it were limited in its capacity to function in the world. Therefore, the grand metaphor associated with this sphere is the Will. The Will is based on freedom. It is the essence of our divinity on earth. We can choose to live in harmony with Divine Law or make the choice not to.

Het-Heru

Het-Heru corresponds to the sphere labeled "7" on the Tree of Life. All of the inventions that people come up with are first designed by visualization in the mind. Furthermore, before an invention becomes a physical reality, a blueprint or design of the thing to be created is first developed. This is another example of how the Sages compared the mental processes of man to the construction of a cosmogony. In this act of the creative process, Neter, in the form of Het-Heru, creates designs of the things that will become physical reality. This is why the grand metaphor for this sphere is the "Celestial Designer" or "Divine Artist".

Sebek

Sebek corresponds to the sphere labeled "8" on the Tree of Life. In this act, Neter, in the form of Sebek, defines and details the designs of the seventh sphere. This is related to Sebek's appellation as the "Divine Messenger", and the defining & segregative nature of words. The preceding sphere concentrates on the whole object while this sphere focuses on its individual parts.

Auset

Auset corresponds to the sphere labeled "9" on the Tree of Life. Up to this point, Neter has brought forth its creative faculties, established a set of laws that the physical world will be bound to, and established a means of enforcing those laws. In addition, Neter has established its ability to function in the world, and brought forth its main faculties for designing and detailing the objects to be created.

In this act, the collective energies of the previous spheres are collected and Neter takes on the form of the Divine Mother, Auset, and gives birth to the forms willed by Ausar. These forms become the various objects of the physical world.

Geb

Geb corresponds to the sphere labeled "10" on the Tree of Life and corresponds to the physical objects at the end of the creative process. In this capacity, Geb is known as the Erpau Neter or Inheritor of God. It means that all of the attributes of the preceding spheres are reproduced and housed in the objects of the physical world. As the saying goes, "as above, so below". This reiterates what has been stated thus far. The creator comes into being in the form of the things that it creates.

To elaborate on this further, consider what happens immediately following conception in the human reproductive process. After the sperm and egg unite, a very remarkable process takes place. This single cell begins a process called mitosis in which it starts to divide. The cell first divides into two identical cells. These two cells will each divide to bring the number to four. Once again, each of these four cells will divide, bringing the number to eight.

Despite increasing differentiation, all of the eight cells still contain the genetic information of the first. This means that the first cell comes into being in the form of the eight. The ancients described the creation of the universe in like fashion. This is evident in Chinese cosmology where on page 2 of *The Numerology of the I Ching*, Huang writes,

> "Confucius's Great Treatise on the I Ching says, "In I, there is Tai Chi, Tai Chi generates two primary energies. Two primary energies generate four primary symbols. Four primary symbols generate eight primary gua".

It is this same process that is expressed in the Egyptian literature as described earlier.

The Tai Chi symbol at the top of the diagram on the following page represents the state of unity in Chinese cosmology. It shows the Yin and Yang energy in perfect balance. The void is expressed in the concept known as Wu Chi. In the Egyptian literature, Amen represents the void, and Ausar, the "unity" between Yin and Yang.

In the diagram, Tai Chi generates the two primary energies of Yin and Yang. The Yin symbol is shown as a broken black line, and the Yang symbol as a solid white line. In the Egyptian literature, Ausar differentiates unformed energy/matter into two primary energies corresponding to Shu & Tefnut or Tehuti and Sekert.

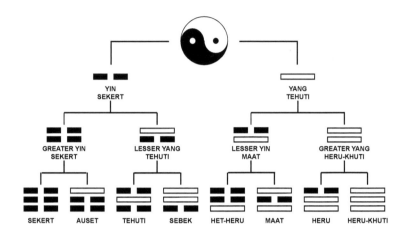

The Eight Primary Gua

Continuing with the image in the diagram, the two primary energies generate four primary symbols by adding a Yin and Yang line to the existing primary symbol. These four symbols correspond to Tehuti, Sekert, Maat, and Heru-Khuti in the Egyptian tradition. To finish with the description of this diagram, the four primary symbols are said to generate 8 primary gua. In Egypt, they correspond to the previous four attributes with the addition of Heru, Het-Heru, Sebek, and Auset. In all, they encompass spheres 2-9 on the Tree of Life.

In relationship to what has just been presented, on page 123 of Anthony T. Browder's *Nile Valley Contributions to Civilization*, he tells us that,

> "This process of halving is the basis of Nile Valley mathematics . . . This concept was expressed in an ancient text which stated: I am One that transforms into Two[,] I am Two that

transforms into Four[,] I am Four that transforms into Eight[,] After this I am One".

This reinforces what has been stated thus far. The eight cells in mitosis are modifications of the first cell. Likewise, the forms of the physical world are modifications of the original state of energy/matter. It is this revelation as expressed in the Egyptian cosmological system that also supports Gann's divisions of wholes into eighths.

THE FOUR ELEMENTS

The four elements are known as Fire, Earth, Water, and Air. They are each associated with two of four qualities that are also used to classify and categorize temperature characteristics in the daily cycle. These four qualities are based on two principles. One is Thermal and the other is Hydrating. The Thermal principal can be dualized into Hot and Cold and the Hydrating principle can be dualized into Dryness and Moisture. In the following sections I will give a brief description of each element, and pair it with its corresponding phase in the daily cycle.

Fire

The element of Fire is classified as both Hot and Dry with heat predominating. Fire produces heat, which gives the molecules of an object additional energy. The energy increases the activity of the molecules so that they bounce against each other, which causes the object to expand. The effect of fire on an object is in harmony with the concept of Ra at midday in the daily cycle. When written without the determinative to denote its divine qualities, ra denotes "work", "action", and "power to do work". It applies directly to the increase in activity experienced by molecules when heated. It is at midday that the earth is exposed to its greatest amount of heat. This is due to the sun's position, which would be more or less, directly overhead. When the sun is directly overhead, the intensity at which its rays strike the earth is at a maximum, thereby causing average daily temperatures to peak.

Earth

The element of Earth is classified as both Cold and Dry with dryness predominating. When there is an increase in the activity of water molecules due to heat, they evaporate. Moisture goes into the air by escaping the object in the form of vapor. This causes the earth to become dry, rigid, and hard. The concept of dryness is in harmony with the time of sunset in the daily cycle. At sunset, the earth is dry due to the intensity of the sun's rays in the preceding phase.

Water

The element of Water is classified as both Cold and Moist with coldness predominating. When an object cools, its molecules contain less energy (heat), which reduces their activity and causes the object to contract. This is in harmony with the concept of Ra and its relationship to midnight in the daily cycle. At midnight, Ra was known as Af, the so-called "dead" sun god, which conveys the idea of being "inactive". It corresponds to a time when the sun's rays do not strike the earth as the sun is below the north horizon. Since the earth is no longer exposed to its primary source of heat (energy), temperatures decrease at this time.

Air

The element of Air is classified as both Hot and Moist with moisture predominating. When objects cool during the night, the layer of air in contact with the object also cools, and causes the condensation of water vapor in that layer of air. Condensation, a chemical reaction in which water is released by the combination of two or more molecules, occurs because the capacity of air to hold water vapor decreases as the air is cooled. Frost forms if the temperature at which condensation begins is below 32°F (0°C). Thus, when an object regains moisture, it becomes flexible and can conform to its surrounding environment.

This corresponds to the last of the key points in the daily cycle, which is sunrise. It is during this period that the earth is just starting to heat up, and therefore, has not lost any of the moisture regained during the night. Some of you may have experienced this moisture in the early morning as the cool and wet touch of grass or the scraping of frost from your car window. This moisture is sometimes called "dew" or "morning dew", a thin film of water that condenses on the surface of objects near the ground.

When we put the four elements and their corresponding temperature and moisture characteristics together, it forms the cycle pictured on the following page. As you can see, the natural order of the elements is dictated by the sun's course during the day. Just as midday turns into sunset, and sunset into midnight, fire turns into earth, and earth into water. By understanding these relationships, it will allow us to effectively categorize the Neteru into one of these eight periods.

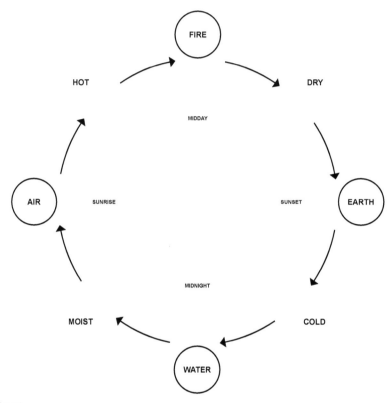

Maat

In ancient Egypt, when depicted in human form, Maat was always depicted as a female. This fact alone gives us a clue to the nature of her energy. In the daily cycle, feminine energy is associated with the period that we call "night". Therefore, we can say that Maat corresponds to a period somewhere after sunset and before sunrise in the daily cycle. To make a finer distinction, all we have to do is look at the symbols associated with this aspect of the Supreme Being.

One of Maat's major symbols is the scales. In the Egyptian judgment scene, the Ab (Will/Heart) of the initiate was weighed against the standard of truth (Ostrich Feather) using Maat's scales. The Ab corresponds to spheres 4 – 6 on the Tree of Life and one of Heru's major symbols. The Ab was placed in one pan of the scale while the ostrich feather was placed in the other. The ostrich feather, also one of Maat's major symbols, represents the standard of truth or Divine Law.

Among other things, the scales represent balance, order, and equilibrium. Within the daily cycle there are only two times when Shu and Tefnut are in balance or equilibrium, they are the times of sunrise and sunset.

If we divide the 24 hours of the day into eight phases, each phase would equal 3 hours in duration. Furthermore, if we say that sunrise occurs at 6 a.m., then its 3-hour phase would span from 4:30 a.m. – 7:30 a.m. Since the basis of Maat's energy is feminine, the beginning of this phase coincides with the time before sunrise (4:30 a.m.). Therefore, taking the whole into consideration, Maat corresponds to sunrise in the daily cycle, which is also the location of the phase labeled "Air".

Interestingly, the symbolism associated with Maat neatly coincides with the characteristics of sunrise in the daily cycle. As the sun rises from below the eastern horizon, there is a point reached within this process where only half of the sun would be visible above the horizon. The remaining half that is not visible would be below the eastern horizon. Therefore, as viewed from the earth, the sun is equally split into two pieces by the horizon line. This is nature's way of showing you the equilibrium corresponding to this time.

Sunrise also gives us another indication that it is associated with the concept of equilibrium. In the course of the daily cycle, sunrise is at the equilibrium between the two extremes of midnight and midday. These facts neatly coincide with the symbolism of Maat, and her major symbol, which is the scales.

In the above diagram, Maat corresponds to the element of Air, which is classified as Hot and Moist. As stated earlier, heat causes objects to expand, and therefore to become light & weightless. This conveys the concept of a rising energy. Moisture causes objects to become flexible. Therefore, it conveys the concept of being able to conform to the surrounding environment. The personality of Maat, as conveyed in the literature, can also be characterized using the qualities described above.

The period between 4:30 a.m. and 7:30 p.m. characterizes a time when the predominantly Yin energy changes into a predominantly Yang energy. Therefore, it can be said that it is expansive, growing, and rising. Its focus is on outward manifestation. If you were to look up words with the same or similar meaning, you will find words such as develop, increase, and prosper. These are metaphors describing the personality characteristics associated with Maat. By living in harmony with Divine Law to which she corresponds, it ensures your success & "prosperity".

In relationship to the moist quality, which conveys the idea of flexibility and accommodation, the personality of Maat can be described as helpful, cooperative, and benevolent. Remember that Divine Law is based on interdependence, a multiplicity of parts working in unison to maintain the whole. It is a cooperative and flexible system.

In addition to the characteristics described above, Maat can be said to possess the mental ability to think systematically or holistically, a function of the right side of the brain. This coincides with the mental ability required for her role in the creative process. Symbolically, she understands how the parts fit into the whole of the universal system.

Heru

In Ancient Egypt, when in human form, Heru was always depicted as a male. Like Maat, this gives us a clue as to the nature of his energy. In the daily cycle, masculine energy is associated with the period that we call "day". Therefore, it can be said that Heru corresponds to the period after sunrise and before sunset in the daily cycle. In order to make a finer distinction as to his place in this cycle, we can look to the literature regarding this attribute of the Supreme Being.

The word hru or heru, written without the determinative to denote its divine qualities, literally means "the day". Heru, like many of his other forms, represented the rising sun. Budge writes on page 486 of *The Gods of the Egyptians Vol. 1* that,

"he represented the new Sun which was born daily".

The new sun would of course correspond to the period after sunrise. Therefore, taking the whole into consideration, Heru corresponds to the period between 7:30 a.m. and 10:30 a.m. within the daily cycle at the phase labeled "Hot".

Heru corresponds to heat, but at a higher degree than Maat. The moisture that was present in the preceding stage has somewhat decreased. Therefore, the increased heat makes Heru's energy impatient, excitable, and forceful. In addition, Heru is characterized as energetic and firm in his movements.

Recall that one of the major symbols of Heru is the Ab. It corresponds to our ability to make an independent decision. To make an independent decision or to truly exercise the will, you have to be full of vitality (Heat). Thus, the hot energy of Heru is the support for those in leadership positions who must make critical decisions. Therefore, it is no surprise that Heru is a major symbol of the kingship in ancient Egypt.

Heru-Khuti

In ancient Egypt, when in human form, Heru-Khuti was also depicted as masculine, and therefore corresponds to the period after sunrise and before sunset. As you can tell by his name, he is also a form of Heru, and hence, the "day". Once again, to make a finer distinction as to what period in the day he corresponds, we can look to the Egyptian literature where we will find a form of Heru-Khuti known as Heru-Behutet.

In regards to this aspect of the Supreme Being, Budge writes on page 473 in *The Gods of the Egyptians Vol. 1* that,

"This is one of the greatest and most important of all the forms of Horus [Heru], for he represents that form of Heru-Khuti which prevailed in the southern heavens at midday and as such typified the greatest power of the heat of the sun".

From the above quoted passage, we learn that Heru-Khuti, in his form as Heru-Behutet, corresponds to the period between 10:30 a.m. and 1:30 p.m. during the daily cycle.

At this position, Heru-Khuti also corresponds to the element of Fire, which is categorized as being both Hot and Dry. In the form of Heru-Behutet, Heru-Khuti was the patron of blacksmiths in the city of Edfu. A blacksmith is one who works directly with fire to forge and shape iron. It is another way of seeing how the Egyptians associated this attribute with this phase in the daily cycle.

Another quality of heat is that it causes separation. For example, if a compound is heated as in distillation, dissimilar substances composing the compound will separate because like substances will cling to their own kind. This concept is evident in the Egyptian literature as it pertains to one of the major symbols associated with Heru-Khuti.

Heru-Khuti is said to enforce the Divine Laws of the 4th sphere. By living in harmony with Divine Law, one invokes Neter's protection. Within our society, it is the police who enforce the law and serve to protect us from harm and danger. To perform this function, they carry and utilize various weapons. Similarly, Heru-Khuti also carries weapons such as knives, spears, and chains. The knife is especially significant, because it is used to cut, divide or separate, which is a quality of heat as described above.

In harmony with the concept of separation is the mental function associated with this aspect of the Supreme Being. The increased heat lends good support for analysis, the 'separation' of an intellectual or material whole into its constituent parts for individual study.

Like Heru, the heat corresponding to this period also makes Heru-Khuti impatient, excitable, and forceful, but to a higher degree. Since the intensity of heat has increased from the Heru phase, the moisture that was responsible for the cooperativeness and sympathy of Maat, has given way to the competitiveness and lack of sympathy of Heru-Khuti. Metaphorically, this is what makes him a perfect candidate to execute judgment and punish those who break the law. He is not one to let personal feelings and sympathy to get in the way.

To carry and utilize dangerous weapons, and to be effective in the heat of battle, it takes a certain level of courage where you don't fear for your life. The support for this courageous behavior is the Hot & Dry metabolic state. In addition to the combative role, heat is also responsible for the competitive nature of athletes. In sporting events, spectators may chant, "Lets get fired up", and before athletes are ready to compete, they warm up. These are key indicators showing us that in order to compete effectively the raising of fire is a necessity.

Sebek

In ancient Egypt, when in human form, Sebek was sometimes depicted with the head of a crocodile, and at others depicted with the head of a jackal. In both instances, the body was depicted as masculine, and therefore corresponds to the period after sunrise and before sunset. To make a finer distinction as to what period in the day Sebek corresponds, we can look to the information regarding one of his various forms.

When depicted with the head of a jackal, Sebek was known as Anpu, who was responsible for embalming the body of Ausar. On page 262 of *The Gods of the Egyptians Vol. 2*, Budge writes,

"Tradition declared that Anubis [Anpu] embalmed the body of Osiris [Ausar] . . . and it was believed that his work was so thoroughly well performed . . . that it resisted the influences of time and decay".

It is the work of the embalmer to preserve the body, and one of the keys to preventing the body from decay is to remove its moisture. It is well known that dry environments such as the desert have preserved bodies from decay for thousands of years.

Taking the whole into consideration, Sebek corresponds to the period between 1:30 p.m. and 4:30 p.m. during the daily cycle. After the sun has reached its peak at midday and temperatures have reached their highs, the earth begins to dry out. Sebek corresponds to the category labeled "Dry", which is positioned in between midday and sunset in the daily cycle.

Dryness possesses the quality of rigidity, which allows an object to define its own form. This is in harmony with Sebek's association with words and their defining nature.

Tehuti

In ancient Egypt, when in human form, Tehuti is depicted with the head of an ibis and body of a male. Therefore, he corresponds to the "day". As there is only one phase remaining in day period, he corresponds to the time of sunset, which covers the period between 4:30 p.m. and 7:30 p.m. in the daily cycle at the phase labeled "Earth".

Despite using the process of elimination to determine Tehuti's position, there is more evidence linking Tehuti to sunset in the daily cycle. On page 411 in *The Gods of the Egyptians Vol. 1*, Budge tells us that,

> "In the Pyramid Texts there is evidence that Thoth [Tehuti]
> was connected with the western sky . . . and this idea is am-
> plified in an interesting fashion in the clxxxth Chapter of the
> Book of the Dead [Book of Becoming Awake], where we find
> that the deceased [initiate] addresses Thoth [Tehuti] both as
> Thoth [Tehuti] and as Temu, the setting sun, or god of the
> west".

Temu is one of the 4 forms of Ra and corresponds to the same
phase in the daily cycle.

Sunset, like sunrise, corresponds to a time when the Yin
and Yang energy are in equilibrium. Once again, the horizon line
splits the sun into two equal pieces. Interestingly, Tehuti is also
associated with the concept of equilibrium in the Egyptian literature. In the story of Ausar, Heru fights his evil uncle Set in order to
avenge the murder of his father. In regards to Tehuti's role in this
story, Budge writes on page 405 in *The Gods of the Egyptians Vol.
1* that,

> "his duty was to prevent either god from gaining a decisive
> victory, and from destroying the other; in fact, he had to keep
> these hostile forces in exact equilibrium".

Tehuti solves the problem of their struggle by finding the point of
equilibrium so that the two opposing forces can coexist with each
other, but in a way that maintains order.

In regards to the element corresponding to this phase, the
Earth is classified as both Cold and Dry. Cold associates dissimilar
substances allowing them to mix together. As stated, dryness refers to rigidity, allowing an object to define its own shape. The earth
is by nature cold because it receives its warmth from exposure to

the sun. It is also dry, and hence rigid, because it possesses a definite shape and form. Its qualities appropriately classify the things that correspond to this phase.

In the area of personality, Tehuti shares many qualities in common with Sebek. Where moisture creates a flexible, creative attitude, dryness creates an inflexible, by the book mode of behavior. It's the type of personality that can be described as dull and boring. Unlike heat, which makes Sebek impatient, fast, and quick thinking, the cold makes Tehuti a little more patient and receptive. This ties into his role as a disseminator of wisdom. To receive counsel from within or from an oracle, you have to become calm so that you are receptive to receive the message.

Auset

When in human form, Auset was depicted as feminine, and therefore corresponds to the period that we call "night". To make a finer distinction, we can look to her role in the creative process as a clue. In the cosmology, we know that Auset is said to give birth to the forms of the physical world, and is thus called the Divine Mother. The motherly energy can best be described as cold, and therefore corresponds to the period between 7:30 p.m. and 10:30 p.m..

Motherly energy can be described as caring and protective. It indicates the type of personality to which she corresponds. Auset conceives and gives birth to the child Heru, but before this occurs she devotes herself to successfully finding the body of Ausar despite the overwhelming adversity she must overcome. She is therefore constant, dedicated, and loyal. Also associated with this phase is the accumulation of water to which she corresponds. It is reflective of her association with trance and the sea.

Sekert

In Ancient Egypt, Seker is depicted as a male, and is wrapped in a long white cloth like that of Ausar. It conveys the fact that they are both mummified, and hence dead. This matches the description of the form of Ra known as Af. Taking this into consideration, Seker corresponds to the same period in the daily cycle at the phase labeled "Water".

The name Seker can also be rendered Seker(t), which gives it a feminine connotation. This further verifies that the symbol belongs to the night period. The tight wrapping of the cloth to which Sekert and Ausar are bound shows that they are inactive. This is the result of the cold quality corresponding to this phase. Like Auset, Sekert is patient, hard working, constant, and steadfast. Where the extreme cold makes Sekert lethargic and slow to act, it also supports the ability for deep protracted thinking.

Another symbol that corresponds to both Sekert and Ausar is Ptah, who was also depicted in the tight fitting garment indicating that he is in mummified form. Recall that it is through the interaction of Tehuti and Sekert that Ausar creates the world. In harmony with this concept, Ptah was said to carry into effect the commands issued by Tehuti concerning the creation of the universe. It is another way of saying that Ptah and Tehuti interacted to carry out the work of creation. This is why you will find Ptah and Sekert merged into one symbol. On page 503 in *The Gods of the Egyptians Vol. 1*, Budge writes that,

> "Ptah-Seker represents a personification of the union of the primeval creative power with a form of the inert powers of darkness, or in other words, Ptah-Seker is a form of Ausar, that is to say, of the night sun, or dead Sun-god".

Once again, we see the reference to the night sun, which of course corresponds to midnight. Ptah represents the creative powers of the 1st and 3rd spheres of the Tree of Life, which is combined into one symbol. Therefore, not only do you find the name Ptah-Seker in the literature, but you will also find "Ptah-Seker-Ausar".

Het-Heru

In ancient Egypt, when in human form, Het-Heru was depicted as feminine, and therefore corresponds to the period after sunset and before sunrise. As there is only one remaining phase to which she can be assigned, this is the period between 1:30 a.m. – 4:30 a.m. in the daily cycle, and corresponds to the phase labeled "Moist".

In the Egyptian literature, it is Auset who conceives and gives birth to the child Heru, but it is Het-Heru, in the form of Nebt-Het, who is said to gestate the child. This ties directly into her association with moisture. On page 29 in the Ageless Wisdom Guide to Healing Vol. 2, Ra Un Nefer Amen writes,

> "Without water the earth cannot produce; fertility in the female (to a lesser extent in males) bears a direct ratio to body fat content (moisture); we speak of "fertile imagination", - these are all integrally related to each other".

Creativity, or the ability to produce, directly corresponds to Het-Heru's role in the creative process as described earlier. Recall that she is the inventive, artistic faculty, and that all inventions are first visualized or imagined in the mind. She also is characterized as sociable, outgoing, and full of joy.

When putting all of this information together, we get the following relationships in the diagram below.

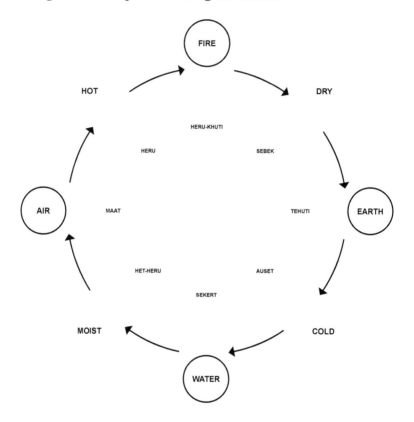

Students of Astrology should be able to easily relate the planets to these divisions of a circle - Auset to the Moon, Seker(t) to Saturn, Het-Heru to Venus, Maat to Jupiter, Heru to the Sun, Heru-khuti to Mars, and Sebek to Mercury. It is my personal belief that all cyclical phenomena exhibit the same eight phases. The four elements, the Neteru, and the cycles associated with them can be utilized as a guide to understanding the phases of any cyclical period. What we learn about one can assist in helping us to learn about another. It helps us to read the signs of the time. For example, the yearly cycle can easily be categorized using what we know so far. The Spring Equinox corresponds to Sunrise, which in turn corresponds to Maat, whose major symbolism is equilibrium. Likewise, the Summer Solstice corresponds to Midday and Fire, the Fall Equinox to Sunset and Earth, and the Winter Solstice to Midnight and Water. So let's look at another cycle and see how its activities coincide with these divisions.

In the female, the reproductive cycle is based on the time elapsed from one menstruation to the next. On average, it is approximately 29.5 days long. During the course of this 29.5 day cycle, there are two key points. The first is the point at which the female begins ovulation, and the second is the point at which she begins menstruation. By understanding what takes place during these two phases, it will aid us in properly categorizing this cycle on the previous diagrams.

At ovulation, the ovaries release a mature egg that will travel into the fallopian tube. There, it will spend anywhere from 12 to 36 hours traveling towards the uterus. It is only during this period that fertilization can occur. That is, if a male sperm cell is able to unite with the female egg. If fertilization does not occur, approximately halfway from the time of ovulation, menstruation will occur.

At menstruation, the nutrient rich lining of the uterine wall that was prepared for a possible embryo will begin to shed. In addition, it can be said that the unfertilized egg that was released into the fallopian tube dies, and preparation will begin for the release of a new egg for the next ovulation.

With the preceding information in mind, we can now match the two key points in the menstrual cycle to its corresponding points in the previous diagrams. As temperature has been a vital part of its content thus far, it will also serve as a vital key for properly categorizing the menstrual cycle.

It is a well-known fact that during ovulation there is a corresponding temperature change that occurs in the female body. It actually rises above its normal average. The relationship between the increase in temperature and the time of ovulation is so distinct that one method for helping individuals who are trying to conceive is through the measurement of body temperature. Each morning, around the time of ovulation, the female will take her body temperature to find the most favorable day to conceive.

The phase labeled "Fire" corresponds to the above average temperatures corresponding to midday in the daily cycle and the summer solstice in the yearly. This is in harmony with the increase in body temperature experienced by a female when she begins to ovulate. Therefore, through association, ovulation corresponds to the phase labeled "fire".

Knowing that menstruation occurs approximately halfway from the time of ovulation, it corresponds to the phase labeled "Water. Through association, menstruation corresponds to midnight in the daily cycle, the winter solstice in the yearly cycle, and to Sekert. Another interesting fact that supports the above relationships is that during menstruation, the female is known to retain water, a fact that further puts it in harmony with the water phase. As mentioned previously, heat in the day causes objects to lose moisture through evaporation, but water accumulates around objects that cool during the night. Therefore, it can also be said that these objects retain water (moisture).

Last but not least, there is one additional fact about the time of menstruation that further puts it in harmony at this point in the cycle. If the egg that was released during ovulation is unfertilized, it is during menstruation that it can be said that the unfertilized egg dies. Interestingly, the faculty corresponding to the water phase is Sekert or Saturn who is said to govern death. Also recall that it corresponds to the form of Ra known as Af, the "dead", inactive aspect of the Life Force.

In summary, Newlands found that every eighth element had similar physical and chemical properties when they were arranged in increasing order of their relative masses. In addition, we found that this could be valid up to calcium (20) only; as beyond calcium (20), elements do not obey the rules of Octaves. Thus, Newlands' Octaves were valid for lighter elements only. Furthermore, it was suggested that the arrangement of elements in Newlands' Octave resembled the musical notes, but this is not the case. It is the seven elements between two elements of similar properties that resemble the seven musical notes.

Elements with Similar Properties	Elements in Between	Notes
Lithium (3) -	- - -	-
	Beryllium (4)	C
	Boron (5)	D
	Carbon (6)	E
	Nitrogen (7)	F
	Oxygen (8)	G
	Fluorine (9)	A
	Neon (10)	B
Sodium (11) -	- - -	-

In reality, there are eight unique elements in this series (3-10), with the ninth (11) bearing similar properties to the first (3).

We found that there was a mathematical formula and sequence of numbers that explains the relationship for all the elements on the periodic table. From this, it was suggested that Gann may have instead derived his theory of eighths from this sequence of numbers and the periodic law. In Gann's *Master Calculator for Weekly Time Periods to Determine the Trend of Stocks and Commodities* course, there is a section with the heading, "DIVISIONS OF TIME PERIODS." Under this section, he shows how to divide the year into 8ths and 3rds and what the equivalency would be in weeks. To the right of the 1/2 or 4/8ths division, there is a comment saying, "a most important time and resistance level". The only other comment to the right of a division is the 3/4ths division or 6/8ths. The comment says, "very important for change in trend." It is worth mentioning that this 3/4ths is 39 weeks, which is 273 days. This falls into the time period that was identified as the evil 6th period when dividing the year into seven periods. It makes me wonder if he understood this, and thus the reason for the special comment under this division.

It is also in this course where we will find information on how Gann divided time periods into eighths as well. Gann writes,

"We divide the cycles into 1/2, which is the most important, and also into the periods of 1/8, 1/3 and 2/3, and watch these proportionate parts of the cycles for changes in trend. For example:

The Great Cycle of 90 years equals 1080 months;
    1/2 is 45 years or 540 months
    1/4 is 22-1/2 years or 270 months
    1/8 is 11-1/4 years or 135 months
    1/16 is 5-5/8 years or 67 1/2 months

The 30-year Cycle or any other cycle is divided up in the same way."

With that said, we still have strong evidence that Gann divided cycles into seven periods in addition to eight. I have already provided examples from the 1919 article as well as his book entitled, *Face Facts America! Looking Ahead to 1950* for the division of the year into seven periods. Now I would like to offer a another perspective. Enter Lyman E. Stowe.

First and foremost, W.D. Gann had an established connection to Lyman E. Stowe. In the e-book entitled, *W.D. Gann on the Law of Vibration*, there was an image of the cover of an issue of the Astrological Bullentina which contained a list of the council members. There are many recognizable names, as W. D. Gann is under Gemini, but what I would like to point out is that Lyman E. Stowe is listed under Aries. Here they are both listed on the council of The Astrological Society, Inc. N.Y.

THE ASTROLOGICAL SOCIETY, INC., N. Y.
ZODIAC COUNCIL.

Office, 1629 Lexington Ave., New York, N. Y.

Meeting first Wednesday of each month at 8 P. M., at 1629 Lexington avenue, New York.

Aries—John Cuadrado, treasurer; J. B. Sullivan, F. A. S.; G. H. Oswald, W. G. Moir. Lyman E. Stowe.

Taurus—Myles McCarthy, M. J. Dix, Mrs. Lohr, Mrs. Campbell, Mrs. F. Wieland, Mrs. J. Sanders.

Gemini—Sir J. Hazelrigg, V. P., F. A. S.; W. D. Gann, Mrs. M. Lindon, Miss G. O'Neil. E. M. Mackey, Mrs. Carter Woods.

Lyman Stowe wrote several books in his day, but the one that I would like to bring to your attention is entitled, *Astrological Periodicity*. Starting on page 276 Stowe writes,

> "I have before stated that every person's good days are the day of the week of birth and the 3d and 4th days of the week after birth. The evil days are the 6th and 7th days after the day of the week of birth. . . The 6th and 7th days of the week have been called good and evil days, since time immemorial. No doubt the basis for the idea of the 6th and 7th days after birth being evil days is that the Bible says God finished his work on the 6th and rested on the 7th days of the week."

We find that this passage is very consistent with what was presented in the early part of this book - Buchanan's discussion of the evil 6th period along with Gann's mention of the 7th as a time of panicky declines. In fact, Mr. Stowe also mentions Dr. Buchanan in his book. On page 279 he writes.

> "Dr. Buchanan figures the evil years of man to come every seven years. But, this does not exactly apply, as we find the regular occurrence of evil years are not exactly in the sevens, yet as a whole they do. They occur between 5 and 6, 13 and 14, 19 and 21, 27 and 28, 33 and 34, 41 and 43, 49 and 50, 54 and 55, 61 and 63, 69 and 70, 75 and 76, 83 and 84, 89 and 90. The reason of this variation is due to regular, planetary influences. The effect of good or evil years will be varied by the cycle one is transiting at the time."

Is it possible that Gann, having a connection with Mr. Stowe, also adopted this theory, that evil years coming every seven years does not exactly apply due to planetary influences. On page 280 of *Astrological Periodicity*, Stowe writes,

> "The Jupiter years are said to be the best years of a persons
> life time. These are the years when Jupiter returns to the po-
> sition it occupied at birth, which occurs every 12 years, or
> between 11 and 12, 23 and 24, 35 and 36, 47 and 49, 59 and
> 60, 71 and 72."

In this example, it is quite easy to identify the good years as it is based on Jupiter returning to the same position at birth. However, the evil years must be based on a number of factors because if we adopt the same procedure, Saturn returns to the same position after 29 years, and the numerous years that he provides don't tally with Saturn's cycle of return. Now, Dr. Buchanan goes on to present a very unique system to detail how certain cycles influence an individual based on when that person was born. He goes on to explain THE CYCLE CHART, which is divided into 84 sections, where there are 7 divisions to one sign. This image is depicted on the following page. On page 291 it reads,

> "This chart has the years of birth from 1 to 84 years of age.
> Starting with Aries and running backward through the signs
> Pisces, Aquarius, Capricornus and so on to the 85th year
> starts life over again. As the years by the 12 signs keep re-
> peating themselves down the left through the signs, which
> stand as a seven year cycle each, the influence of each year is
> repeated except as is changed through the cycle it is in. . . The
> years and cycles can be calculated from this chart . . ."

THE CYCLE CHART

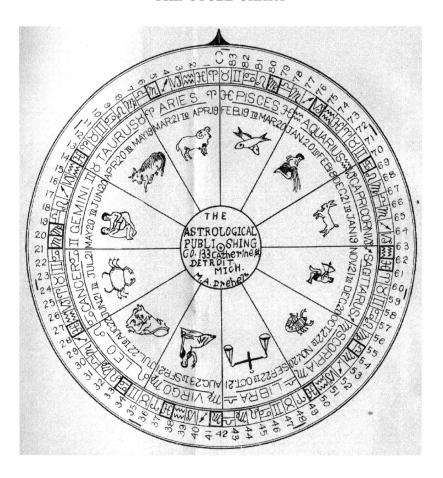

I have often quoted Gann and his references to the use of mathematics. In an interesting passage on page 293 of *Astrological Periodicity*, Stowe writes,

> "Kabalistic Astrology is based upon mathematics, as calculated with planetary influences. The ancients understood this thoroughly; as no man has been able to add to or take from one preposition of Euclid, they depend, much on figures. The ancients divided the planets into octaves, as ruling the harmony of vibrations . . ."

I wonder if these zones of influence have certain planetary associations as I had pointed out with the division into eighths, and that while transiting in a certain zone, whether it be time or price, an individual or stock is sensitive to the planet corresponding to that zone or sensitive in a certain way to other planets based on the zone being transited. This is just food for thought and will require more research. It is here that I will bring this small booklet to a close. I hope that you will find something of value in these pages to continue your own research.

BIBLIOGRAPHY

Amen, Ra Un Nefer, *Ageless Wisdom Guide to Healing Vol. 2: Health Teachings of the Ageless Wisdom*, Khamit Corporation, Bronx, NY 1983

Barrett, W. F., "Light and Sound: An Examination of Their Reputed Analogy." *The Quarterly Journal of Science.* Vol. VII. London: Longmans, Green, and Co., Paternoster Row. January 1870. p. 1-16.

Browder, Anthony T., *Nile Valley Contributions to Civilization: Exploding the Myths Vol. 1*, The Institute of Karmic Guidance, Washington, D.C., 1992

Buchanan, Jos. Rhodes, *Periodicity: The Absolute Law of the Entire Universe.* Chicago: A. F. Seward & CO., 1897.

Budge, E. A. Wallis, *The Gods of the Egyptians Vol. 1: Studies in Egyptian Mythology*, Dover Publications, Inc., Mineola, NY, 1969

Budge, E. A. Wallis, *The Gods of the Egyptians Vol. 2: Studies in Egyptian Mythology*, Dover Publications, Inc., Mineola, NY, 1969

Duncan, Robert Kennedy, *The New Knowledge: A Simple Exposition of the New Physics and the New Chemistry in Their Relation to the New Theory of Matter*. New York: A. S. Barnes & Company, 1906

Fagan, Cyril, *Astrological Origins*, Llewellyn Publications, St. Paul, Minnesota, 1971

Fishbough, William, *The End of the Ages: With Forecasts of the Approaching Political, Social and Religious Reconstruction of America and the World*. New York: Continental Publishing Company, 1898.

Gann, W.D., *Face Facts America! Looking Ahead to 1950*. New York: W. D. GANN & SON, Inc., 1940

Gann, W.D., *Master Calculator for Weekly Time Periods to Determine the Trend of Stocks and Commodities*. January 10, 1955

Gann, W.D., "Sees the Kaiser Shot While Trying to Flee His Prison." *The Milwaukee Sentinel Magazine*. 5 January 1919.

Gann, W.D., *The Tunnel Thru the Air Or Looking Back From 1940*. New York: Financial Guardian Publishing Co., 1927.

Huang, Alfred, *The Numerology of the I Ching: A Source book of Symbols, Structures, and Traditional Wisdom, Inner Traditions*, Rochester, Vermont, 2000.

Newlands, John A. R., *On the Discovery of the Periodic Law, and On Relations Among the Atomic Weights*. New York: E. & F. N. Spon, 1884.

Orlock, Carol, *Know Your Body Clock*, Barnes & Noble, Inc., New York, 1993

Schucht, J. H., "Number Seven a Law of Nature" *English Mechanic And World of Science*, No. 1711. January 7, 1898, p. 484.

Schucht, J. H., "The Physical Basis of Music" *English Mechanic And World of Science*, No. 1320. July 11, 1890, p. 425.

Stowe, Lyman E., *Astrological Periodicity*, Astrological Publishing Co., Detroit, Mich., 1909

Who Was OROLO?. Gann Study Group. http://finance.groups.yahoo.com/groupgannstudygroup/.

Wyckoff, Richard D., "William D. Gann: An Operator Whose Science and Ability Place Him in the Front Rank - His Remarkable Predictions and Trading Record." *The Ticker and Investment Digest*, Vol. 5, No. 2. December, 1909: 51-55.

ALSO AVAILALE FROM THE AUTHOR

W.D. GANN: DIVINATION BY MATHEMATICS

W.D. GANN: DIVINATION BY MATHEMATICS:
HARMONIC ANALYSIS

Made in the USA
Columbia, SC
25 April 2021